D'ZUELE GESCHICHT

THE NUMBER STORY

SMALL BOOK ONE
ENGLISH - LUXEMBOURGISH

Numbers Teach Children
Their Number Names

written and illustrated by

MISS ANNA

Early Reader Edition of *The Number Story 1*
Bronze Medal Winner, 2016 Wishing Shelf Book Award

Library of Congress Control Number: 2018902040

Names: Miss Anna, author.
Title: Number story : numbers teach children their number names / Miss Anna.
Description: Portland, OR: Lumpy Publishing, 2018.
Identifiers: ISBN 978-1-949320-03-9 | LCCN 2018902040
Summary: The pictures and rhymes present stories which introduce numbers 0-10.
Subjects: LCSH Numeration—English—Luxembourgish--Pictorial works--Juvenile literature. |
BISAC JUVENILE NONFICTION /
Languages: English—Luxembourgish
Classification: LCC QA141.3 .M57 2018 | DDC 513—dc23

Publisher: Lumpy Publishing
Website: www.missannabooks.com
Email: missanna@missannabooks.com

Paperback: ISBN 978-1-949320-03-9
Printed in the U.S.A.   1 3 5 7 9 10 8 6 4 2

Want to learn our number names?

Wéilst du gäre d'Nimm vun den Zuelen léieren?

It is very easy and a lot of fun!

Et ass ganz einfach and mëcht och vill Spass!

Say-along our little jingle

Sang éis't klëngt Geschicht mat éis!

starting from Number One!

Mär fänken mat der Nummer Eent un!

1

ONE looks like my one finger.

EENT

gesäit aus wéi main Fanger.

1
ONE!
EENT!

2

TWO   trails a tail.

ZWEE

huet een Schwanz.

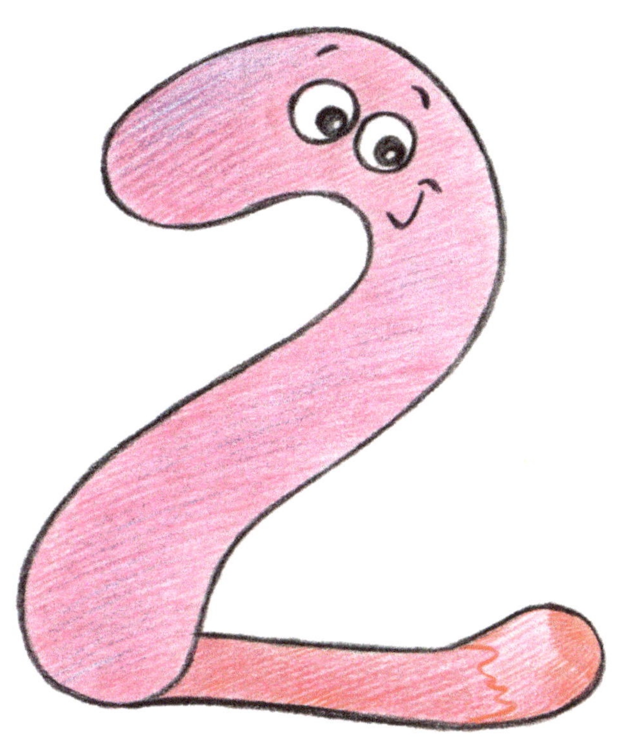

A TAIL!   EEN SCHWANZ!

3

THREE has bumps.

DRÄI

huet knuppe.

BUMPY! KNUPPËCH!

4

FOUR   carries a sail.

VÉIER

huet e Segel.

4
A SAIL!
E SEGEL!

5
FIVE   is a racing track.
FËNNEF
as eng Rënnstrëck.

VROOM
VROOOM!

6

SIX curves like a snail.

SECHS

as gekurvt wéi e Schleek.

A SNAIL! E SCHLEEK!

7

SEVEN has a sharp angle.

SIWEN

huet e spatzen Wënkel.

BE CAREFUL! IT'S SHARP!

Opgepasst! Et as Spatz!

8

EIGHT is rollercoaster rails.

AACHT

as eng Aachterbunn.

YAY!
YIPPEE!

NINE   is a bubble on a stick.

NÉNG

as eng Bloos op engem Stéck.

A BUBBLE!

ENG BLOOS!

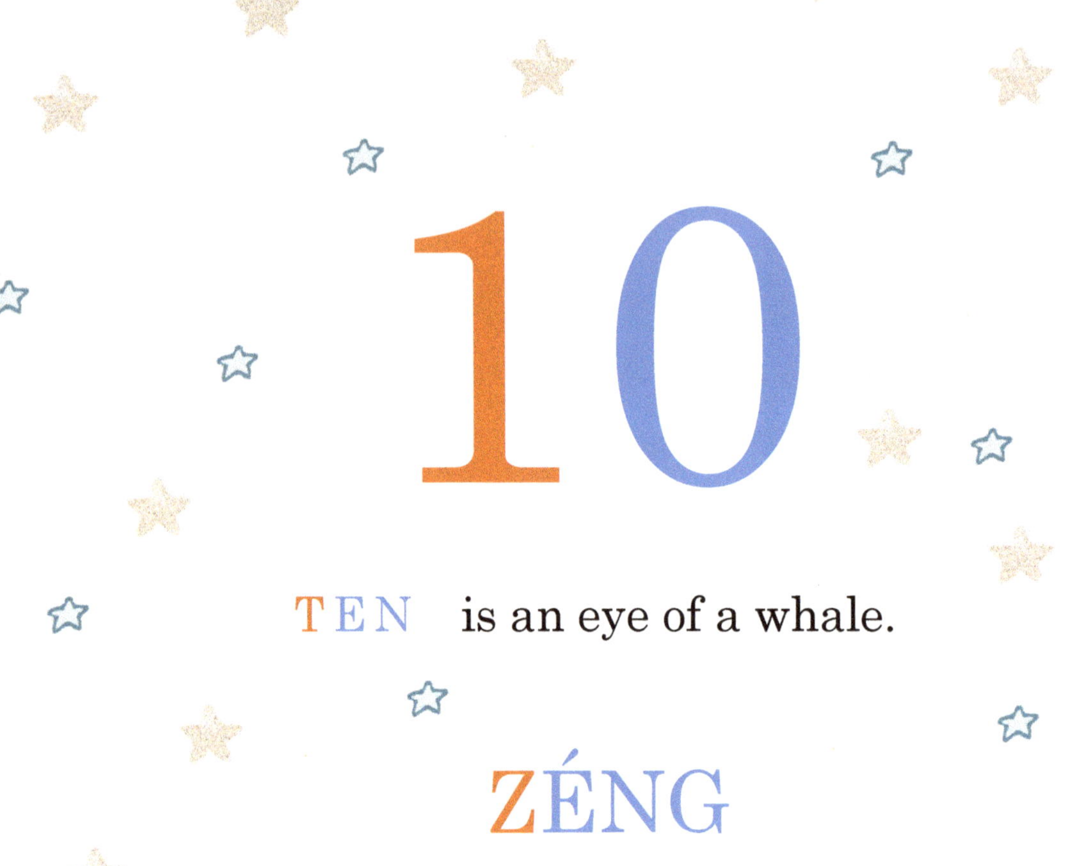

10

TEN   is an eye of a whale.

ZÉNG

as een Aen vun engem Waal.

WINK!
ZWINKERE!

HELLO! HALLO!

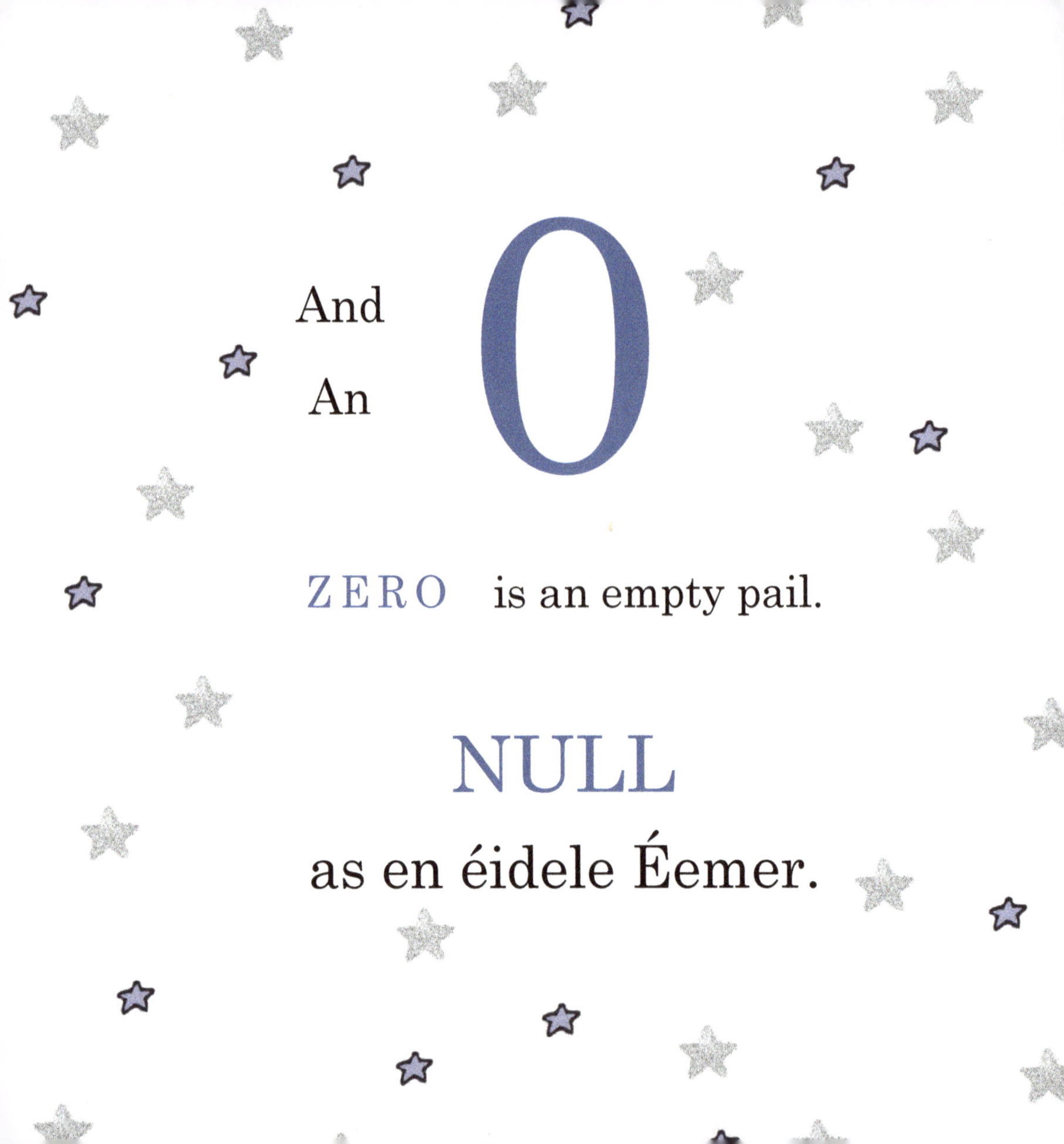

And

An

0

ZERO is an empty pail.

NULL

as en éidele Éemer.

IT'S EMPTY!
EN ASS ÉIDEL!

Thank you for playing with us today.

We had a lot of fun too!

Merci daatse haut mat eis gespillt hues.

Mir haaten och immens vill Spass!

We are your Number friends,
Zero to Ten,
Who will be here for you~

Mir sin deng Zuele Frënn
Null bis Zéng.

Mir wäerte ëmmer fir dëch do sin.

Bye-bye now!
See you again soon!

Äddi äddi!

Bis geschwën!

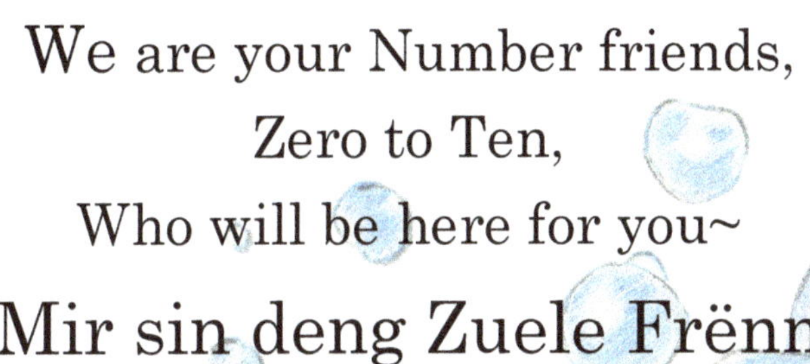

The Numbers are *SINGING* too!

To sing-a-long, look for Miss Anna Number Story
at your favorite music store like iTUNES.

MP3

Numbers 0-10
IDENTIFYING
& COUNTING

Numbers 11-20
& Ordinals

first, second, third...

Numbers 0-100
& Place Values

ones, tens, hundreds...

About Clocks
& Telling Time

hours, minutes, seconds

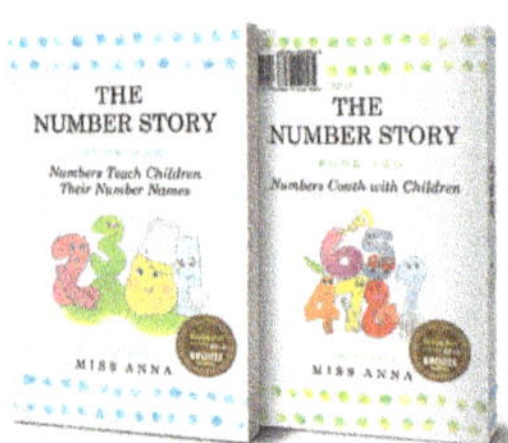

Number Story 1 & 2

isbn: 978-0-996216-48-7

Number Story 3 & 4

isbn: 978-1-945977-01-5

Number Story 5 & 6

isbn: 978-1-945977-06-0

Number Story 7 & 8

isbn: 978-1-949320-40-4

For more Miss Anna books to love,
visit us at

www.missannabooks.com

Numbers are working hard all over the world!
Come Travel the World with Us!